ABRAMS TANKS

BY DENNY VON FINN

EPIC

BELLWETHER MEDIA · MINNEAPOLIS, MN

EPIC BOOKS are no ordinary books. They burst with intense action, high-speed heroics, and shadows of the unknown. Are you ready for an Epic adventure?

This edition first published in 2013 by Bellwether Media, Inc.

No part of this publication may be reproduced in whole or in part without written permission of the publisher. For information regarding permission, write to Bellwether Media, Inc., Attention: Permissions Department, 5357 Penn Avenue South, Minneapolis, MN 55419.

Library of Congress Cataloging-in-Publication Data

Von Finn, Denny.
  Abrams tanks / by Denny Von Finn.
      p. cm. – (Epic books: military vehicles)
  Includes bibliographical references and index.
  Audience: Grades 2-7.
  Summary: "Engaging images accompany information about Abrams tanks. The combination of high-interest subject matter and light text is intended for students in grades 2 through 7"–Provided by publisher.
  ISBN 978-1-60014-814-9 (hbk. : alk. paper)
  1. M1 (Tank)–Juvenile literature. 2. Tanks (Military science)–Juvenile literature. I. Title.
  UG446.5.V65 2013
  623.74'752–dc23   2012002389

Printed in the United States of America, North Mankato, MN.

TABLE OF CONTENTS

Level

Latitude > 44°

Longitude > 20°

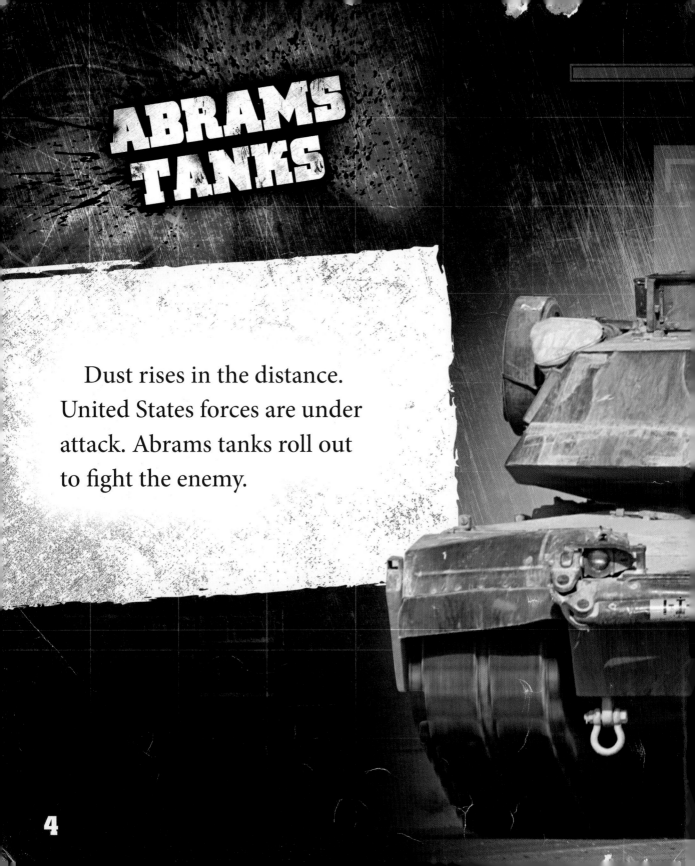

ABRAMS TANKS

Dust rises in the distance. United States forces are under attack. Abrams tanks roll out to fight the enemy.

Latitude > 44° 49' 6" N

Longitude > 20° 28' 5" E

B-43

F4CAV

5

Rounds shoot out of the large Abrams guns. Enemy tanks explode in the distance. They are no match for the mighty Abrams!

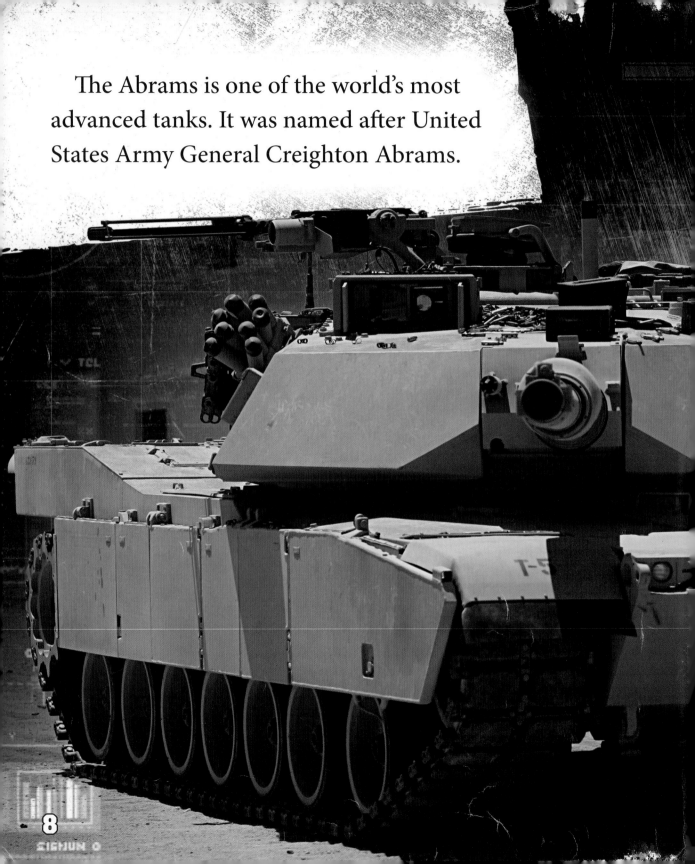

The Abrams is one of the world's most advanced tanks. It was named after United States Army General Creighton Abrams.

Abrams Fact

Creighton Abrams
served in World War II,
the Korean War, and
the Vietnam War.

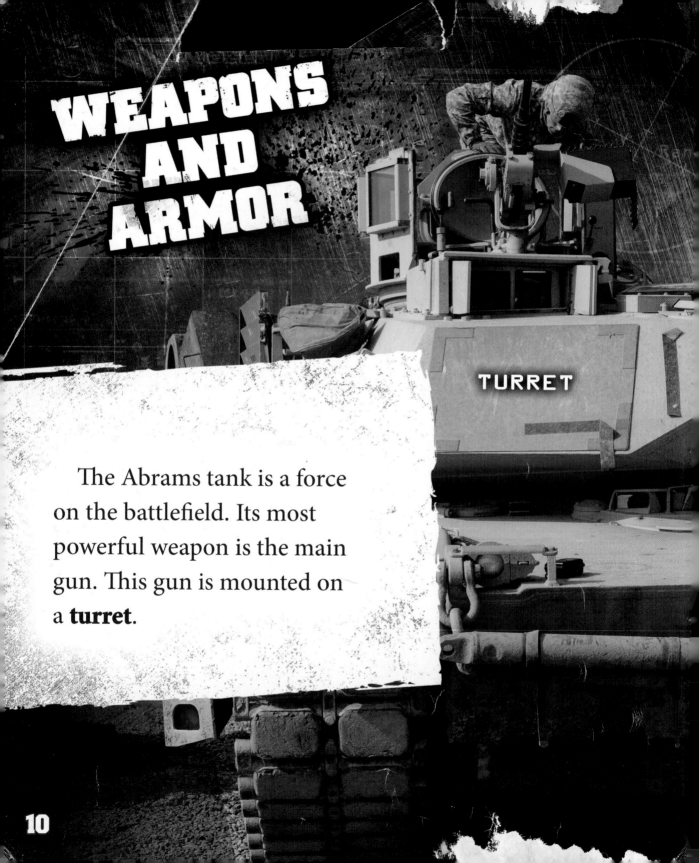

WEAPONS AND ARMOR

TURRET

The Abrams tank is a force on the battlefield. Its most powerful weapon is the main gun. This gun is mounted on a **turret**.

MAIN GUN

The **gunner** turns the turret to aim the gun. The **commander** and gunner use computers to find targets. They work with the driver and **loader** to complete **missions**.

GUNNER

COMMANDER

Abrams tanks have strong **armor** to protect their crews. This armor is up to 24 inches (61 centimeters) thick. It stops most enemy fire.

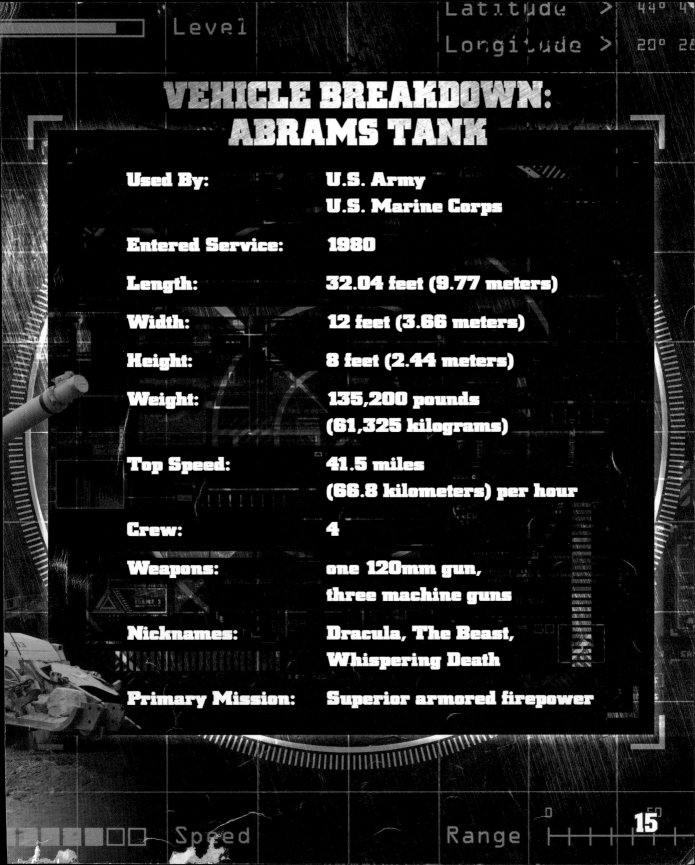

VEHICLE BREAKDOWN: ABRAMS TANK

Used By:	U.S. Army U.S. Marine Corps
Entered Service:	1980
Length:	32.04 feet (9.77 meters)
Width:	12 feet (3.66 meters)
Height:	8 feet (2.44 meters)
Weight:	135,200 pounds (61,325 kilograms)
Top Speed:	41.5 miles (66.8 kilometers) per hour
Crew:	4
Weapons:	one 120mm gun, three machine guns
Nicknames:	Dracula, The Beast, Whispering Death
Primary Mission:	Superior armored firepower

Speed

Range

Latitude > 44° 4
Longitude > 20° 2

Level

15

ABRAMS MISSIONS

313-69▲

065

TRACKS

The Abrams tank is made to move quickly and surprise the enemy. **Tracks** help it move over almost any kind of land. Crews use the tank's firepower to destroy enemy vehicles.

Abrams Fact

An Abrams tank can climb over objects more than 4 feet (1.2 meters) tall!

The Abrams first saw battle
in the **Gulf War**. Commanders
used **periscopes** to find the
enemy. Loaders kept the tanks
ready to fire.

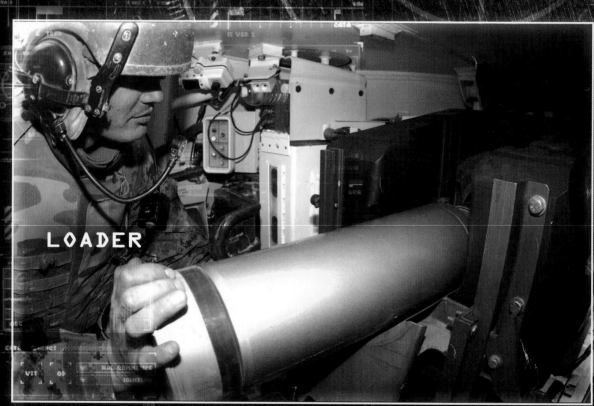

LOADER

PERISCOPE

The Abrams tank proved its power in the Gulf War. It continues to rule battlefields today.

GLOSSARY

armor—thick plates that cover the tank to protect the crew; Abrams armor is made of layers of steel, ceramic, and plastic.

commander—the Abrams crew member who directs the gunner, loader, and driver

Gulf War—a conflict from 1990 to 1991 in which 34 nations fought against Iraq; the war began after Iraq invaded the small country of Kuwait.

gunner—the Abrams crew member who searches for targets and fires the tank's guns

loader—the Abrams crew member who places rounds into the main gun

missions—military tasks

periscopes—telescope-like devices that stick out of the top of tanks; tank commanders and drivers use periscopes to see their surroundings.

rounds—ammunition; each round has all of the parts needed to fire one shot.

tracks—long belts that wrap around the wheels of an Abrams tank; tracks help an Abrams grip the land.

turret—the part of the tank that rotates and holds the main gun

TO LEARN MORE

At the Library

David, Jack. *Abrams Tanks*. Minneapolis, Minn.: Bellwether Media, 2008.

Jackson, Kay. *Military Tanks in Action*. New York, N.Y.: PowerKids Press, 2009.

Loveless, Antony. *Tank Warfare*. New York, N.Y.: Crabtree Pub. Co., 2008.

On the Web

Learning more about Abrams tanks
is as easy as 1, 2, 3.

1. Go to www.factsurfer.com.

2. Enter "Abrams tanks" into the search box.

3. Click the "Surf" button and you will see a list
of related Web sites.

With factsurfer.com, finding more information
is just a click away.

INDEX